Keep Your Students in Mathematical Shape!

Mathercise™
Classroom Warm-Up Exercises

Book E
For Advanced Algebra, Pre-Calculus, Third or Fourth-Year High School Math

Michael Serra

Playing It Smart
P.O. Box 27540
San Francisco, CA 94127

mserramath@gmail.com
www.michaelserra.net

Printed in the United States of America
10 9 8 7 6 5 4 3 2 1   15 14 13 12
ISBN: 978-1-55953-063-7

What is *Mathercise E?*

Mathercise E is a mathematical fitness program designed to keep students who are taking advanced algebra, pre-calculus, or a third or fourth-year high school math course in great mathematical shape. It is a series of 50 class starters which strengthen important math skills. You can easily integrate *Mathercise E* into your existing math program because it requires very little additional preparation. Instead of taking up valuable class time, a regular *Mathercise* program will actually give you more time to teach because it encourages your students to come to class on time and immediately get "on task."

The book contains 50 reproducible *Mathercise Book E* masters. You can use the masters to create overhead transparencies or reproduce them as individual student worksheets. Each *Mathercise* is a set of three problems and a space for a fourth. The first problem (Reason) is always an inductive or deductive reasoning problem. The second problem (Solve) is a symbolic thinking problem that reviews one of a number of basic mathematical skills (averages, percent, probability, algebra, proportional thinking, functions, and coordinate geometry). The third problem (Sketch) is a visual thinking problem that requires students to think in two and three dimensions and graph linear, absolute value, quadratic, and exponential functions — many with horizontal and/or vertical shifts.

In the first problem the student must use inductive or deductive reasoning.

The second problem asks the student to solve a proportion, averages, percent, or probability problem. Students also solve systems of equations for one or more variables and evaluate compositions of functions and "fantasy functions."

The third problem requires the student to sketch a 2- or 3-dimensional figure, usually transformed in some way. Students are also asked to graph a variety of functions, including trigonometric functions.

In the blank space for the fourth problem, you, the teacher, are to write a review problem from an earlier class or homework assignment.

MATHERCISE E50

1. Reason.

From the given set of clues, find a 3-digit number that satisfies all the clues.

674 pico
032 pico
519 fermi
349 bagels
560 pico
???

2. Solve.
A bag contains 4 red, 3 blue, and 2 green marbles. Two marbles are taken from the bag. What is the probability that the 2 marbles are green?

3. Sketch.
Sketch one period of the graph of $y = \cos\left(x - \frac{\pi}{4}\right) + 1$.

4. Review.

What are the objectives of *Mathercise E?*

- *Mathercise* gets students in the habit of being "on task" right from the start of class. All too often the first 5 to 10 minutes of class time is taken up with routine clerical activities. Students learn that it is smart to be tardy. *Mathercise* can encourage more produce study habits and promote punctuality.

- *Mathercise* give students the opportunity to develop and practice both inductive and deductive reasoning skills. Math is supposed to be the subject where students develop reasoning skills. However, we rarely give students an opportunity to build and practice these skills before demanding that they apply them to new topics. *Mathercise* lets students develop reasoning skills in a familiar context, without the added burden of learning new mathematical content.

- *Mathercise Book E* gives students needed practice with the basic skills of proportional thinking, probability, and the graphing of linear, absolute value, and quadratic functions.

- *Mathercise Book E* gives students practice drawing 2- and 3-dimensional geometric figures. Many students have difficulty drawing basic geometric shapes. This useful skill can be learned with practice.

- *Mathercise* provides you, the teacher, with an alternate means of motivating students to do their homework and a quick method of checking on student progress. With busy schedules and large classes it is nearly impossible to look over all the homework of all of our students. *Mathercise* can be used as an alternative or additional check on student progress. By placing an assigned homework problem in the review slot (problem 4) of a *Mathercise*, you can encourage students to do their homework. If students are permitted to use their notebooks for the *Mathercise*, it even prompts them to keep accurate notes.

- *Mathercise* prepares students for college entrance exams. The second problem on each *Mathercise* is similar to one type of problem appearing on standardized tests. You can use the review problem to practice additional problem types found on standardized tests.

- *Mathercise* once a week will keep your students in a great mathematical shape!

How do you use *Mathercise Book E* in your classroom?

Use *Mathercise* as a class starter, getting students "on task" immediately at the start of the class period. *Mathercise* is designed to be used once or twice a week with students who have taken a year of algebra and are taking advanced algebra, pre-calculus, or a third or fourth-year high school math course. Here are some suggestions that will help you use *Mathercise Book E* effectively.

- Make transparencies from the masters for use with an overhead projector. If you can't use an overhead projector, make individual copies for each student in the class. Students shouldn't write on the copies, but instead should put their answers on a separate sheet of paper. For the third problem (Sketch) require students to sketch the figure at a different size. This will ensure that they don't merely trace.

- In the blank space (problem 4) of a *Mathercise* transparency, or on a copy of a master, write an actual problem from a past homework assignment. Or, write a problem of the type found on college entrance exams, or any other type of skill you wish to have your students practice such as estimation or graphing.

- Make available to students, as they enter class, answer strips of paper large enough to work all four problems (create your own or use the answer sheet masters provided with *Mathercise Book E*).

- Immediately following the first bell (signaling the beginning of the passing period between classes), place the *Mathercise* transparency on the overhead projector or pass out individual copies.

- Give students 10–15 minutes after the second bell to work all four problems. Meanwhile, take attendance, return papers, or attend to other classroom chores. You can also use this time to circulate and look at homework.

- Encourage students to work with pencil and straightedge, carefully sketching what they see on the *Mathercise* master.

- Let your students correct their own papers as you demonstrate solutions. Or, ask students to trade papers and correct their neighbor's work. You may find that you need to correct the sketch (problem 3) yourself.

- There are a variety of ways to use *Mathercise*. At the beginning of the year you may wish to use *Mathercise* daily to get students used to the idea of coming to class on time, and then move to a once-a-week schedule. Another alternative is to use each *Mathercise* for three days. For example, on Monday your students could work problem 1 (Reason) and problem 4 (Review). On Wednesday, students could work problem 2 (Solve) and a new problem 4. On Friday your students could work problem 3 (Sketch) and another problem 4.

Tips for Doing *Mathercise* Problems

Mathercise problems, while not routine, are self-explanatory enough to not require a lot of teaching. They become gradually more challenging and students learn to do them with practice. Initially, though, you'll want to give students some tips for how to solve different types of problems that they may not have much experience with. What follows are tips for some problem types found in problems 1 and 3. You can use these examples to teach these problem types when students first encounter them.

Tips for Problem 1 (Reason)

The first problem in each *Mathercise* (Reason) is an inductive or deductive reasoning problem. In inductive reasoning problems the task is fairly straightforward: the student must find the next term in a number or picture pattern. There are, however, a number of different types of deductive reasoning problems. You should work the following examples with your students to familiarize them with these types of problems.

Following Directions

Some students may be unfamiliar with North, South, East, and West. Show students examples of Northwest, Southwest, Northeast, and Southeast. You should also review with them *right, left,* and *about face* turns. A turn to the right or left means a turn of 90°. An about face means a turn of 180°. The best approach to solving this type of problem is to teach students to move their pencils through the steps of the problem (let their fingers do the walking). This problem type is first encountered in Mathercises E5 and E7.

Example

- **Problem:** You are facing west. You turn right, then about face, and then left. Which direction is to your right?
- **Solution:**

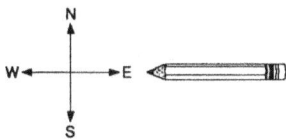

You are facing west.   You turn right,   then about face,   then left.

The pencil is now facing east, so the directions to your right is south.

Verbal Reasoning

Many students need to be convinced that it is okay to read and reread a problem over and over again. You should model the reasoning necessary to solve this problem type, by reading and rereading the statements slowly and carefully working your way backwards through the problem. This problem type is first encountered in Mathercises E6 and E8.

Example

- **Problem:** Print the word that is repeated three times in this sentence and underline the vowels in the word.
- **Solution:** The first task is to find a word that is repeated three times in the sentence. The word is *the*. What do we do with this word? Underline the vowels. Therefore, the answer is th<u>e</u>.

Logical Ranking

Drawing diagrams in an organized way helps to solve this type of problem. Model the reasoning necessary to solve this problem type. I recommend that you begin by drawing your translation of the first statement. Then you should draw a new improved ranking based on the second sentence. Continue in this manner until the problem is solved. This problem type is first encountered in Mathercises E9 and E15.

Example

- **Problem:** Annie is funnier than Bonnie. Cara is funnier than Donna. Cara is not as funny as Bonnie. Which one is the funniest? Who is the least funny?

- **Solution:**

ANNIE	ANNIE		CARA		ANNIE
BONNIE	BONNIE	?	DONNA		BONNIE
					CARA
					DONNA

Annie is funnier than Bonnie.

Cara is funnier than Donna.

Cara is not as funny as Bonnie.

Logical Matching

Visual thinking is very helpful in solving this type of problem. To model the reasoning necessary to solve this problem type, I recommend that you begin by making a chart with names on one vertical side and occupations across the top. Then systematically cross out boxes that are logically eliminated and place circles in boxes that are logically matched. This problem type is first encountered in Mathercises E18 and E20.

Example

- **Problem:** Ace, Borus, and Cathy have careers as airplane mechanic, pilot, and air traffic controller but not necessarily in that order. The airplane mechanic works with Ace. Cathy hired the pilot to fly her to Mexico. Borus earns less than the airplane mechanic, but more than the air traffic controller. Match the names and occupations.

- **Solution:**

	Mech	Pilot	Controller
Ace			
Borus			
Cathy			

	M	P	C
A	X		
B			
C		X	

	M	P	C
A	X		
B	X		
C	O	X	X

	M	P	C
A	X	X	O
B	X	O	X
C	O	X	X

Ace, Borus, and Cathy have careers as airplane mechanic, pilot, and air traffic controller.

The airplane mechanic works with Ace. Cathy hired the pilot to fly her to Mexico.

Borus earns less than the airplane mechanic,

but more than the air traffic controller.

Therefore Ace is the air traffic controller, Borus the pilot, and Cathy the mechanic.

Reasonable 'Rithmetic

To model the reasoning necessary to solve this problem type, I recommend that you begin by finding what letter you can first determine. Then rewrite the problem replacing the solved letter with the number. This problem type is first found in Mathercises E25 and E27.

Example

- **Problem:** Each of the letters in the sum on the right represents a different digit. What is the value of A?

$$\begin{array}{r} A\ B \\ +\ B\ A \\ \hline 1\ B\ 5 \end{array}$$

- **Solution:** The first thing to notice is that $A + B$ adds up to 15 and not just 5, because $A + B$ is also in the tens column. Therefore, in the tens column, $A + B +$ (the carry of 1) gives a total of 16. Thus $B = 6$ and $A = 9$.

Bagels

Bagels began as a logical guessing game on computers. In a bagels puzzle you are to determine a 3-digit number (no digit repeated) by taking "educated guesses." After each guess you are given a clue about your guess. The clues:

> **bagels**: No digit is correct.
> **pico**: One digit is correct but in the wrong position.
> **fermi**: One digit is correct and in the correct position.

In each bagels problem, a number of guesses have been made with the clue for each guess shown to the right. From the given set of clues, you can determine the 3-digit number.

Because of space limitations, the meaning of the clues bagels, pico, and fermi will not appear with the *Mathercise* bagels problems. Therefore, you should make a poster of the rules for bagels and post it for use throughout the year (or at the very least, write it on the chalkboard on the day it appears so your students will have it while they work). To model the reasoning necessary to solve this problem type, I recommend that you begin by making a list of all the digits 0 through 9 and then cross out all those that can be eliminated. Note that if a digit from 0 through 9 has not been eliminated, it can appear in the answer, even if it doesn't appear in the problem. This problem type is first encountered in Mathercises E26 and E28.

Example

- **Problem:**

  ```
  234 pico
  567 pico
  891 fermi
  641 bagels
  825 pico
  ???
  ```

- **Solution:**

2 3 4̸ P	2 3 4̸ P	2 3 4̸ P	②3 4̸ P
5 6̸ 7 P	5 6̸ 7 P	5̸ 6̸ 7 P	5̸ 6̸ ⑦ P
8 9 1̸ F	8 9 1̸ F	8̸ ⑨1̸ F	8̸ ⑨1̸ F
6̸ 4̸ 1̸ B	6̸ 4̸ 1̸ B	6̸ 4̸ 1̸ B	6̸ 4̸ 1̸ B
8 2 5 P	8 2 5 P	8̸ 2 5̸ P	8̸ ②5̸ P
▢ ▢ ▢	▢ 9 ▢	▢ 9 ▢	7 9 2

The first thing to notice is that 1, 4, and 6 are eliminated (from line 4, 641 = bagels). Therefore, one of the numbers is either a 2 or a 3, but not both. One of the numbers is either a 5 or a 7, but not both. And one of the numbers is either an 8 or 9, but not both.

We can eliminate the 8 because if 8 is correct, then it must be in the first or left hand position (line 3 = fermi). But line 5 has a pico with an 8 in the first or left most position. This is a contradiction. If 8 is eliminated and 1 is eliminated, then 9 must be correct in the center position.

If 8 is eliminated, then in line 5 either 2 or 5 is correct but in the wrong position. 5 can be eliminated because if 5 is correct, then it must be in the center position (line 2 says 5 cannot be in first position and line 5 says 5 cannot be in the third position). But this is a contradiction because 9 must be in the center position.

Therefore 5 is eliminated and 2 must be correct in the third position. If 5 and 6 are eliminated, then 7 is left for the first position and thus our solution is 792.

Tips for Problem 2 (Solve)

The second problem (Solve) in each *Mathercise* is a symbolic thinking problem. By *symbolic thinking* we mean the problem usually requires the use of algebra or the ability to use symbols to solve it. Each problem reviews one of a number of basic mathematical skills (averages, percent, probability, algebra, proportional thinking, functions, and coordinate geometry). Many of these problems can be found in college entrance exams such as the SAT. Your students should be familiar with most of the types of problems found in the problem 2 portion of the *Mathercise*, except perhaps the problems we call *Fantasy Functions*. These problems appear quite often in SAT exams.

Fantasy Functions

Most students will be unfamiliar with the concept of creating new binary operations. You should show your students examples of new fantasy functions and how to evaluate them for given values. In *Mathercise Book E* this problem type is first encountered in Mathercises E8 and E21.

Example

- **Problem:** If $a \Delta b = a^2 - 2b$, and $f \Delta 5 = f + 32$, where $f > 0$, find f.
- **Solution:** The rule: $a \Delta b = a^2 - 2b$, says, whatever is to the left (a) of the binary operator "Δ" gets squared (a^2) and whatever is to the right (b) of the binary operator "Δ" gets doubled ($2b$). For example, if $a = 3$ and $b = 4$, then $3 \Delta 4 = (3)^2 - 2(4)$ or 1. Therefore, $f \Delta 5 = (f)^2 \Delta 2(5) = f^2 - 10$. But it was given that $f \Delta 5 = f + 32$; therefore $f^2 - 10 = f + 32$. Solving for f we get:

  $f^2 - f - 42 = 0$
  $(f + 6)(f - 7) = 0$
  Therefore, $f = {}^-6$ or 7
  But $f > 0$, therefore $f = 7$

Tips for Problem 3 (Sketch)

The third problem in each *Mathercise* (Sketch) is a visual thinking problem that requires students to think in two and three dimensions and be able to graph linear, absolute value, quadratic, and exponential functions — many with horizontal and/or vertical shifts. You should practice horizontal and vertical shifts of these functions with your students.

Answers for *Mathercise Book E*

Mathercise E1	1. L	2. 23	
Mathercise E2	1. 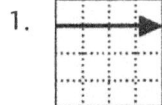	2. 7.5 lbs.	
Mathercise E3	1. 22	2. 50	3. 
Mathercise E4	1. 	2. 12	3. 
Mathercise E5	1. Northwest	2. x	3. 1 cm
Mathercise E6	1. e	2. $y = \sqrt[3]{x-3}$	3. 45°
Mathercise E7	1. Southwest	2. 8	3. See Answer Below
Mathercise E8	1. *stop*	2. 8/15	3. See Answer Below
Mathercise E9	1. G O R Y B	2. 17	3. See Answer Below
Mathercise E10	1. Four	2. $y = \frac{-5}{3}x + \frac{19}{3}$	3. See Answer Below

E7. 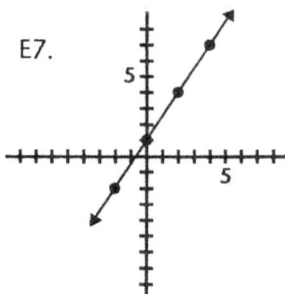   E8. 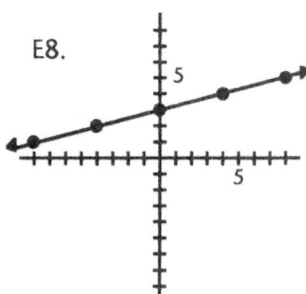   E9. 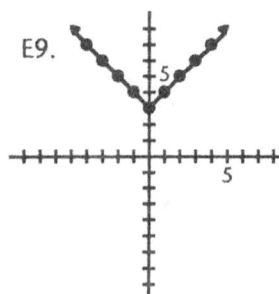   E10.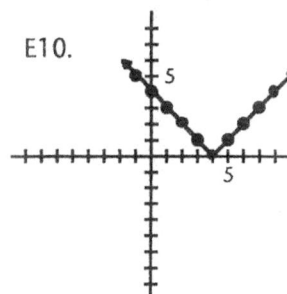

Mathercise E11	1. 387	2. $y = \frac{-2}{3}x - \frac{7}{3}$	
Mathercise E12	1. 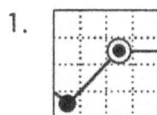	2. $\frac{1}{9}$	
Mathercise E13	1. 131	2. $\frac{1}{3}$	3.
Mathercise E14	1. 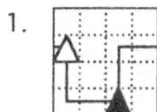	2. 56	3. 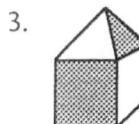
Mathercise E15	1. Richard	2. 5.06	3. 

Mathercise E16	1. s	2. $c = \frac{9}{10}a$	3. 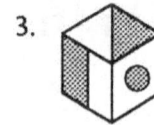
Mathercise E17	1. Ydal Ecnal Arres Dirgni	2. $z = \frac{3}{10}x$	3. See Answer Below
Mathercise E18	1. Anna – Basketball Tomita – Baseball Carla – Swimming Heather – Tennis	2. 9	3. See Answer Below
Mathercise E19	1. Yes, Klinger	2. 2450	3. See Answer Below
Mathercise E20	1. Art – Basketball Tom – Football Carl – Tennis Hector – Swimming	2. $x^2 + 5x + 6$ or $(x + 2)(x + 3)$	3. See Answer Below

E17. 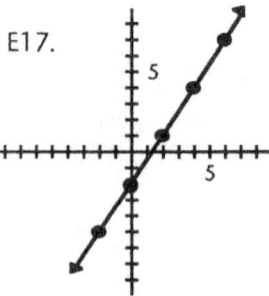   E18. 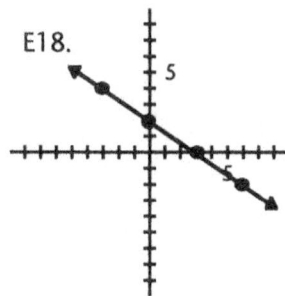   E19. 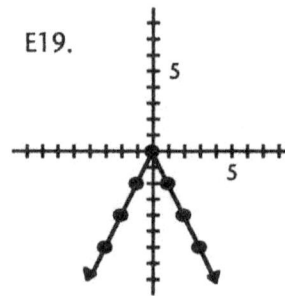   E20.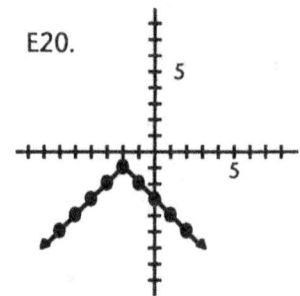

Mathercise E21	1. 96	2. $k = 12$	
Mathercise E22	1. 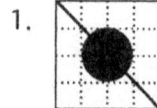	2. $y = \frac{-3}{4}x - \frac{3}{2}$	
Mathercise E23	1. 110	2. $y = 5$	3. 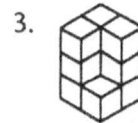
Mathercise E24	1. 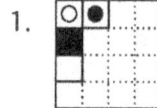	2. 4	3.
Mathercise E25	1. $B = 4$	2. $\frac{1}{3}$	3. 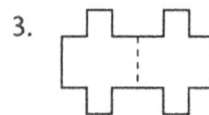
Mathercise E26	1. 358	2. $\frac{17}{36}$	3. 9 minutes
Mathercise E27	1. $B = 1$	2. 15 m	3. See Answer Next Page
Mathercise E28	1. 286	2. $z = 2w$	3. See Answer Next Page

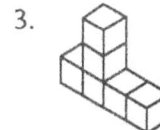

Mathercise E29	1. $C = 4$	2. $z = \frac{1}{10}x$	3. See Answer Below
Mathercise E30	1. 982	2. $x^2 + 2xh + h^2 + 3x + 3h + 2$ or $(x + h + 1)(x + h + 2)$	3. See Answer Below

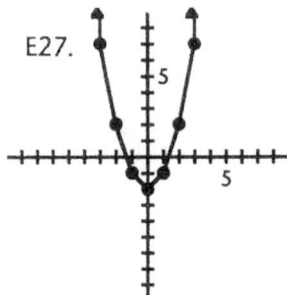

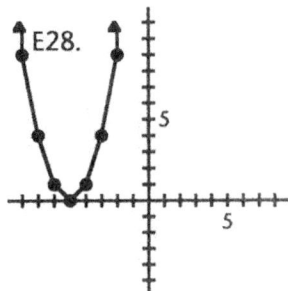

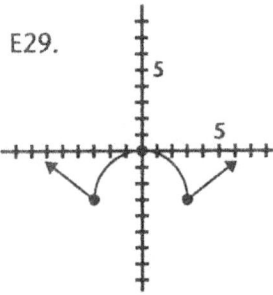

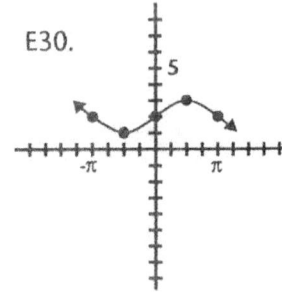

E27.

E28.

E29.

E30.

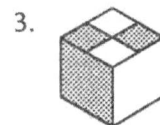

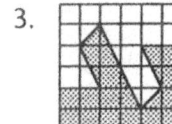

Mathercise E31	1. 3906	2. $2xh + h^2 + 3h$	
Mathercise E32	1.	2. 8	
Mathercise E33	1. 307	2. $k = {}^{-}2$ or 5	3.
Mathercise E34	1.	2. $\sqrt{109}$	3.
Mathercise E35	1. South	2. $y = \frac{3}{10}x$	3. No
Mathercise E36	1. *bale*	2. (0,3)	3.
Mathercise E37	1. 250	2. $\frac{13}{36}$	3. See Answer Next Page
Mathercise E38	1. South	2. $\frac{1}{6}$	3. See Answer Next Page
Mathercise E39	1. Steve – Martha Mark – Susan Robert – Ruby	2. 400	3. See Answer Next Page
Mathercise E40	1. Yes, Rosita	2. $V = 12x$	3. See Answer Next Page
Mathercise E41	1. 300	2. $V = 10y^2$	

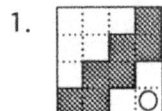

E37.

E38.

E39.

E40.

$-\frac{\pi}{2}$ $\frac{\pi}{2}$ π $\frac{3\pi}{2}$

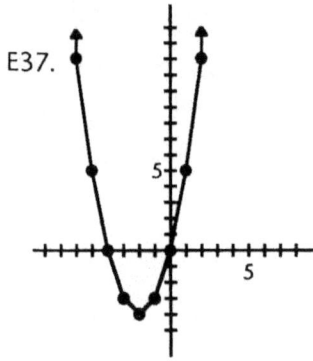

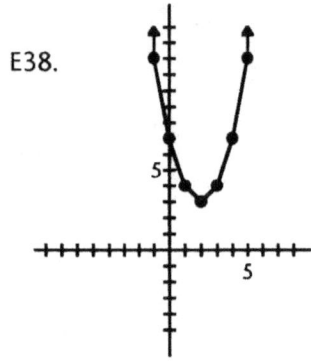

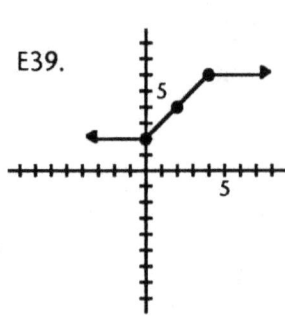

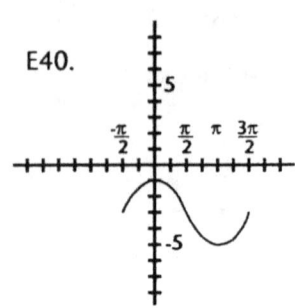

Mathercise E42

1. 23

2. $2x + h + 3$

Mathercise E43

1. 598

2. $3x^2 + 3xh + h^2$

3.

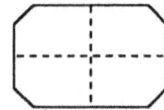

Mathercise E44

1. 99

2. $x = {}^-1$

3.

Mathercise E45

1. Corvette – Harriet
Porsche – Dick
Alfa Romeo – Tom

2. $8Q + 2$ or
$2(4Q + 1)$

3. No

Mathercise E46

1. $A + B + C = 6$

2. $x = -1$

3.

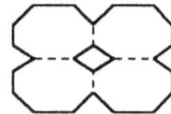

Mathercise E47

1. 417

2. $y = (x - 2)^2 + 4$

3. See Answer Below

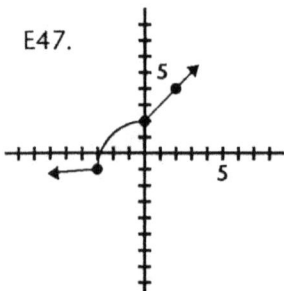

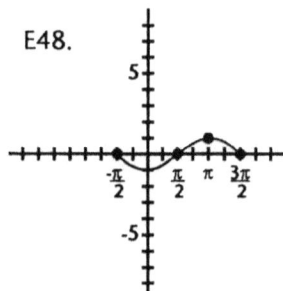

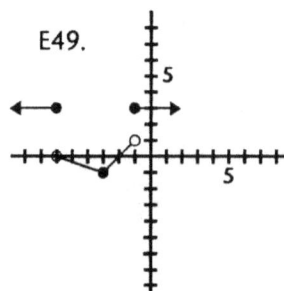

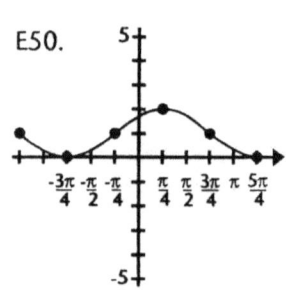

Mathercise E48

1. Spotted Gelding –
Natalie
Black Stallion – Doreen
Brown Mare – Toni

2. $y = \frac{-1}{8}(x + 1)^2 + 2$

3. See Answer Below

Mathercise E49

1. $A = 3$

2. $\frac{1}{12}$

3. See Answer Below

Mathercise E50

1. 216

2. $\frac{1}{36}$

3. See Answer Below

E47.

E48.

$-\frac{\pi}{2}$ $\frac{\pi}{2}$ π $\frac{3\pi}{2}$

E49.

E50.

$-\frac{3\pi}{4}$ $-\frac{\pi}{2}$ $-\frac{\pi}{4}$ $\frac{\pi}{4}$ $\frac{\pi}{2}$ $\frac{3\pi}{4}$ π $\frac{5\pi}{4}$

NAME _____
DATE _____ PERIOD _____
MATHERCISE _____

1.

2.

3.

4.

NAME _____
DATE _____ PERIOD _____
MATHERCISE _____

1.

2.

3.

4.

NAME _____

DATE _____ PERIOD _____

MATHERCISE _____

1.

2.

3.

4.

NAME _____

DATE _____ PERIOD _____

MATHERCISE _____

1.

2.

3.

4.

MATHERCISE E1

1. Reason.

What comes next in the pattern?

Z, 2, X, 3, V, 5, T, 7, R, 11, P, 13, N, 17, __?__ ,

2. Solve.

The average of 4 numbers is 18 and the average of 5 other numbers is 27. What is the average of all 9 numbers?

3. Sketch.

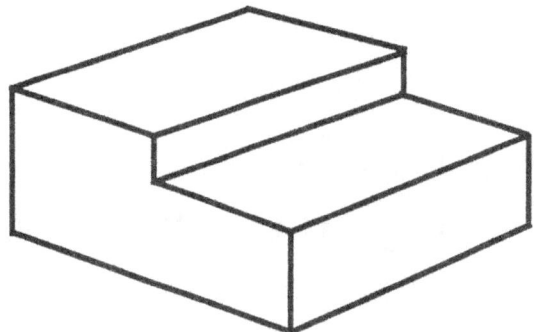

4. Review.

MATHERCISE E2

1. Reason.

If the pattern continues to the right, draw which direction the 99th arrow would point.

2. Solve.

At the last party Al Dente prepared 18 meatballs from 2.5 lbs. of hamburger. Al is planning another party and needs 54 meatballs. How many pounds of hamburger does he need?

3. Sketch.

4. Review.

MATHERCISE E3

1. Reason.

What comes next in the pattern?

5, 10, 3, 8, 0, 5, -4, 1, -9, -4, -15, -10, __?__ ,

2. Solve.

$4x - 5y = 19$
$6x + 7y = 31$

Find $10x + 2y$.

3. Sketch.

Sketch this figure resting on its shaded faces.

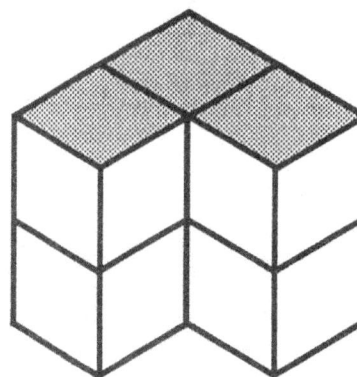

4. Review.

MATHERCISE E4

1. Reason.

If the pattern continues to the right, draw how the 96th box would look.

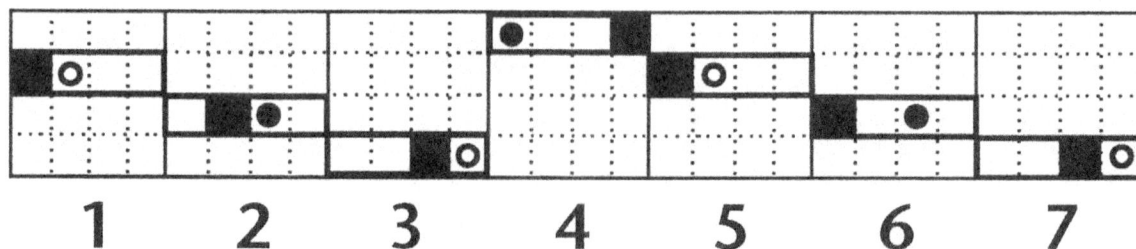

1   2   3   4   5   6   7

2. Solve.

$7x - 8y = 16$

$3x - 6y = 4$

Find $4x - 2y$.

3. Sketch.

Sketch what this shape would look like if it were folded along the dotted lines into a solid figure.

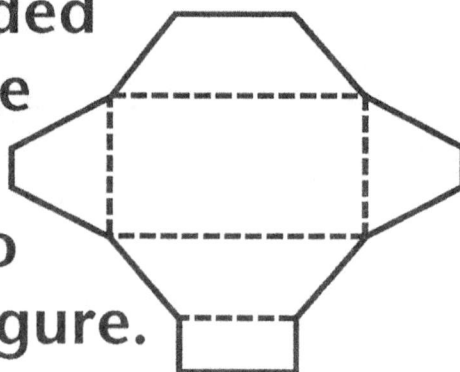

4. Review.

MATHERCISE E5

1. Reason.

You are facing southwest. You turn about face, then left, then about face again. Which direction is directly behind you?

2. Solve.

$f(x) = 2x + 3$

$g(x) = \frac{1}{2}(x - 3)$

Find $f(g(x))$.

3. Sketch.

Rectangle $ABCD$ represents a sheet of paper 21 cm wide. The dotted lines PQ and RS are fold lines. When the 7 cm flap $APQD$ and the 4 cm flap $BRSC$ are folded over on top of the center, what is the width of the overlap?

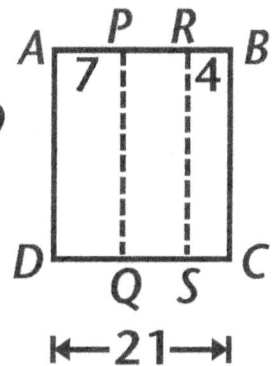

4. Review.

MATHERCISE E6

1. Reason.

What is the letter in the word *television* which is two letters before the vowel that precedes the letter *s*?

2. Solve.

$$f(x) = x^3 + 3$$

Find $f^{-1}(x)$.

3. Sketch.

What is the exact measure of the angle formed by the hands of a clock at 7:30?

4. Review.

MATHERCISE E7

1. Reason.

You are facing northeast. You turn left, then about face, then left again. Which direction is directly behind you?

2. Solve.

$f(x) = x^2 + 3$
$g(x) = \sqrt{x - 3}$

Find $f(1) + \bar{g}^1(1)$.

3. Sketch.

Sketch the graph of $y = \frac{3}{2}x + 1$.

4. Review.

MATHERCISE E8

1. Reason.

If the third letter of the word *Washington* comes before the twentieth letter of the alphabet then print the word *Stop*. Otherwise print the word *Washington* with the last six letters deleted.

2. Solve.

$$a \# b = \frac{1}{a} + \frac{1}{b}$$

Find 3 # 5.

3. Sketch.

Sketch the graph of the line through (0,3) with a slope of $\frac{1}{4}$.

4. Review.

MATHERCISE E9

1. Reason.

There are exactly five houses on a block running East and West. The corner houses have no pets. The orange house is not a corner house. The blue house is the eastern neighbor of the yellow house. The red and yellow houses have pets and are neighbors. The green house is west of the orange house. Draw five adjacent squares in a horizontal column representing the five houses and correctly label them according to their color.

2. Solve.

What is the length of segment *AB* if *A* has coordinates (2, -5) and *B* has coordinates (10,10)?

3. Sketch.

Sketch the graph of the absolute value function $y = |x| + 3$.

4. Review.

MATHERCISE E10

1. Reason.

What is the word in this sentence that has four letters and the second letter is a vowel but the last letter is not e, s, l, or t, however, the last letter is also the sixth letter in the word clever.

2. Solve.

What is the equation in slope-intercept form of the line passing through (2, 3) and (5, -2)?

3. Sketch.

Sketch the graph of the absolute value function $y = |x - 4|$.

4. Review.

MATHERCISE E11

1. Reason.

What comes next in the pattern?

3, 16, 39, 72, 115, 168, 231, 304, ___?___ ,

2. Solve.

What is the equation in slope-intercept form of the line passing through (1, -3) and parallel to the line $y = \frac{-2}{3}x - 4$?

3. Sketch.

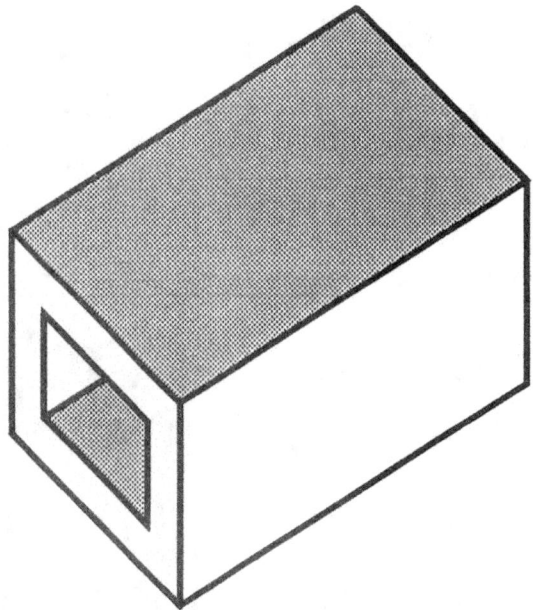

4. Review.

MATHERCISE E12

1. Reason.

If the pattern of circles and connecting segments continues to the right, draw how the 99th box would look.

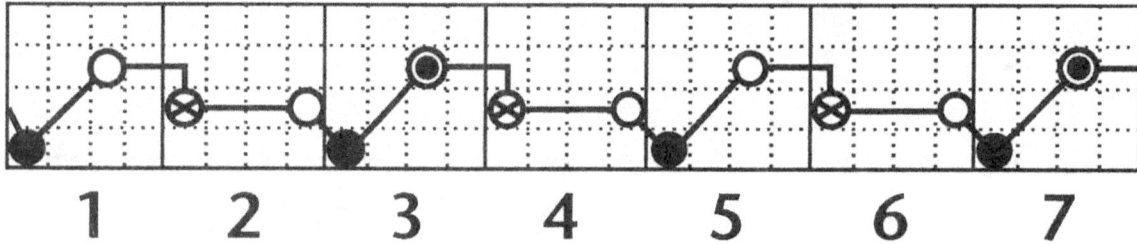

1   2   3   4   5   6   7

2. Solve.

What is the probability of rolling either a sum of 3 or a sum of 11 with a pair of 6-sided dice?

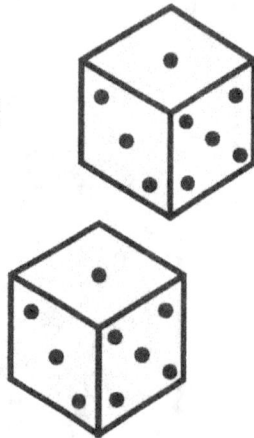

3. Sketch.

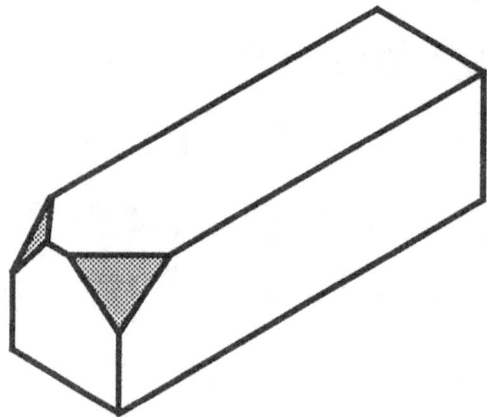

4. Review.

MATHERCISE E13

1. Reason.

What comes next in the pattern?

4, 5, 7, 11, 19, 35, 67, __?__ ,

2. Solve.

What is the probability of rolling a 5 on one or the other of a pair of 6-sided dice?

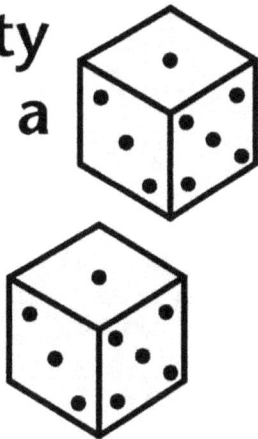

3. Sketch.

Sketch this figure resting on its shaded faces.

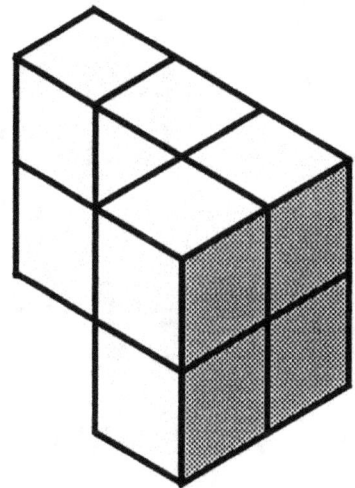

4. Review.

MATHERCISE E14

1. Reason.

If the pattern continues to the right, draw how the 89th box would look.

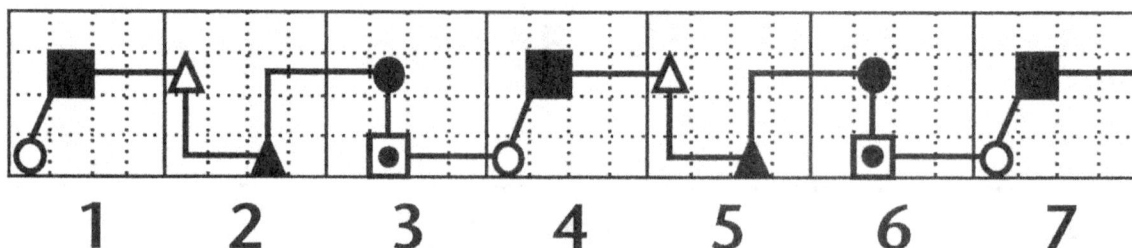

1 2 3 4 5 6 7

2. Solve.

A class of 30 students took a test that was scored from 0 to 100. Twenty-four had scores greater than or equal to 70. What is the lowest possible class average?

3. Sketch.

Sketch what this shape would look like if it were folded along the dotted lines into a solid figure.

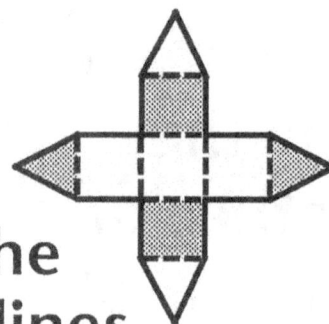

4. Review.

MATHERCISE E15

1. Reason.

May is older than Ramona but younger than Richard. May is older than Ryan who is older than Ramona. Who is the oldest of the four?

2. Solve.

Knuckles McCoy allowed 18 earned runs in 32 innings. What is the pitcher's earned run average? That is, at this rate, how many runs would the pitcher allow in 9 innings? Round your answer to the nearest hundredth.

3. Sketch.

Sketch one of the possible shapes formed when you place the shaded faces of one solid against the second.

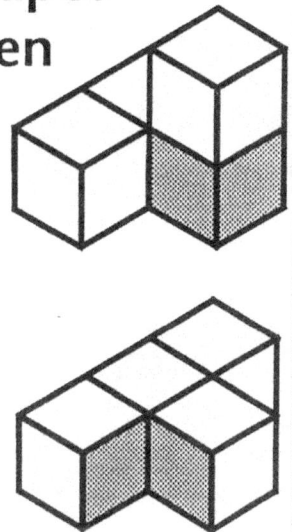

4. Review.

Mathercise Book E

MATHERCISE E16

1. Reason.

If the third letter after the second letter of the word *baker* is the last letter of that word, print the letter in the alphabet that immediately follows that last letter of that word. Otherwise, give yourself a nice compliment by printing the words "I'm ok" within a circle.

2. Solve.

$a = \frac{2b}{3}$

$b = \frac{5c}{3}$

Find c in terms of a.

3. Sketch.

Each pair of opposite faces on the cube below are identical except that the shading is reversed. Sketch a view of the cube showing the three hidden faces.

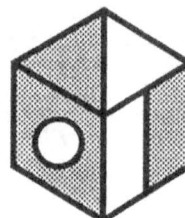

4. Review.

MATHERCISE E17

1. Reason.

Arres is funnier than Dirgni but not as funny as Ecnal. Ecnal is funnier than Dirgni but less so than Ydal. Write the names of the four "things" in a column, from funniest on top to the least funny on the bottom.

2. Solve.

$$x = \frac{y}{5}$$

$$y = 10w$$

$$w = \frac{5z}{3}$$

Find z in terms of x.

3. Sketch.

Sketch the graph of the line with y-intercept $(0, -2)$ and parallel to $y = \frac{3}{2}x + 1$.

4. Review.

MATHERCISE E18

1. Reason.

Anna, Tomita, Carla, and Heather are on four different sports teams. Each girl plays on only one team. They play baseball, basketball, tennis, and swimming. Tomita plays shortstop on her team. The tallest of the four plays basketball and the shortest plays tennis. Carla is taller than Heather but shorter than either Tomita or Anna. Match each person with their sport.

2. Solve.

$f(x) = x + 3$
$g(x) = |x| + 3$

Find $f(g(-3))$.

3. Sketch.

Sketch the graph of the line with x-intercept (3, 0) and perpendicular to $y = \frac{3}{2}x + 1$.

4. Review.

MATHERCISE E19

1. Reason.

Oliver Drab, Beetle Bailey, Sergeant Shultz, and Klinger are professional soldiers. Oliver Drab and Beetle Bailey are the same rank. Beetle Bailey has a lower rank than Sergeant Shultz. Sergeant Shultz has a lower rank than Klinger. From the information given, is it possible to determine whether Oliver Drab is higher or lower ranked than Klinger? If yes, who is ranked higher of the two?

2. Solve.

$$f(x) = x^2 + 3$$

$$g(x) = |x^2| - 3$$

Find $f(-35) + g(-35)$.

3. Sketch.

Sketch the graph of the absolute value function $y = -2|x|$.

4. Review.

MATHERCISE E20

1. Reason.

Art, Tom, Carl, and Hector are on four different sports teams. Each guy plays on only one team. They play football, basketball, tennis, and swimming. Tom's position is quarterback. The tallest of the four plays basketball and the shortest is a swimmer. Carl is taller than Hector but shorter than either Tom or Art. Match each person with their sport.

2. Solve.

$$f(x) = x^2 + 3x + 2$$

Find $f(x + 1)$.

3. Sketch.

Sketch the graph of the absolute value function
$$y = -|x + 2| - 1.$$

4. Review.

MATHERCISE E21

1. Reason.

What comes next in the pattern?

5, 6, 5, 9, 6, 15, 10, 26, 19, 44, 35, 71, 60, 109, ___?___ ,

2. Solve.

$a \# b = \dfrac{1}{a} + \dfrac{1}{b}$

$6 \# k = \dfrac{1}{4}$

Find k.

3. Sketch.

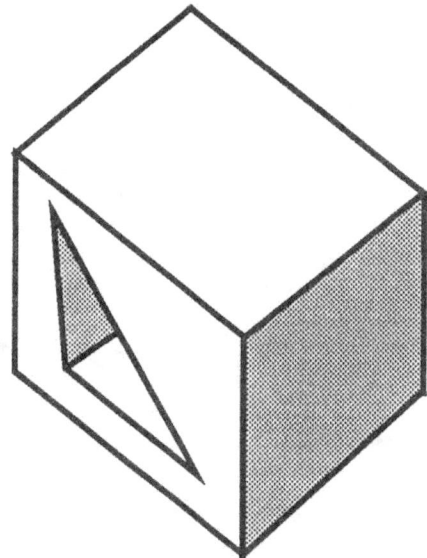

4. Review.

MATHERCISE E22

1. Reason.

If the pattern continues to the right, draw how the 34th box would look.

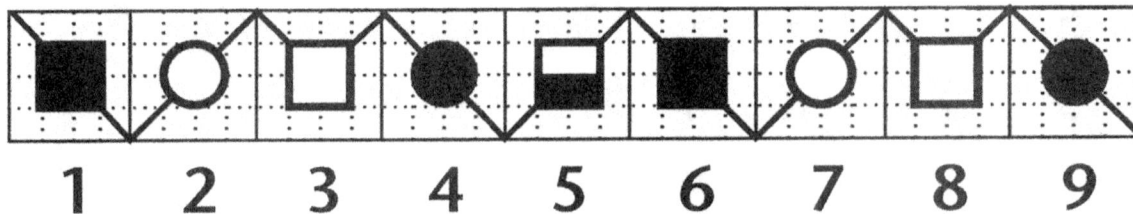

```
  1   2   3   4   5   6   7   8   9
```

2. Solve.

What is the equation in slope-intercept form of the line passing through (2, -3) and parallel to the line $y = \frac{-3}{4} x - 2$?

3. Sketch.

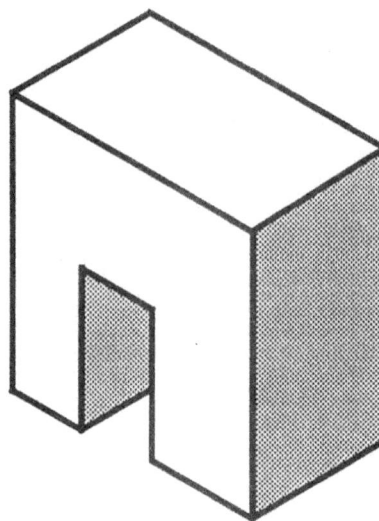

4. Review.

MATHERCISE E23

1. Reason.

What comes next in the pattern?

6, -2, -4, 0, 10, 26, 48, 76, __?__ ,

2. Solve.

What is the equation in slope-intercept form of the line passing through (2, 5) and parallel to the line that passes through the points (0, 0) and (3, 0)?

3. Sketch.

Sketch this figure resting on its shaded faces.

4. Review.

MATHERCISE E24

1. Reason.

If the pattern continues to the right, draw how the 42nd box would look.

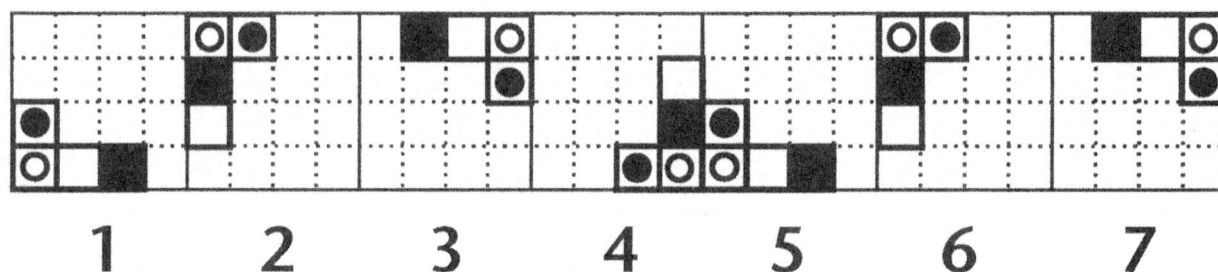

1   2   3   4   5   6   7

2. Solve.

What is the length of segment *MN* if *M* is the midpoint of side *AC* and *N* is the midpoint of side *BC* in triangle *ABC* with vertices *A* at (1, 1), *B* at (9, 1), and *C* at (7, 9)?

3. Sketch.

Sketch what this shape would look like if it were folded along the dotted lines into a solid figure.

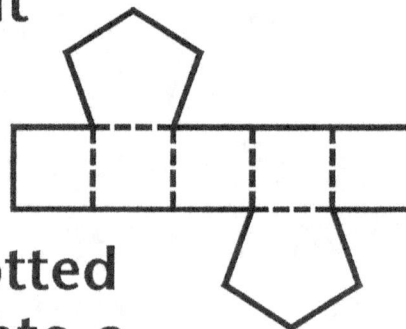

4. Review.

MATHERCISE E25

1. Reason.

Each of the two letters in the product on the right represents a different digit. What is the value of *B*?

$$\begin{array}{r} 3\ B\ 5 \\ \times\qquad A \\ \hline 2\ 0\ 7\ 0 \end{array}$$

2. Solve.

What is the probability of rolling a 6 on one or the other of a pair of 6-sided dice?

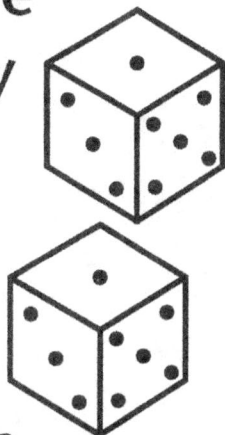

3. Sketch.

An $8\frac{1}{2}$ x 11 sheet of paper is folded in half vertically. Snip off a one-inch square from each corner and sketch what the paper will look like unfolded.

4. Review.

MATHERCISE E26

1. Reason.

From the given set of clues, find a 3-digit number that satisfies all the clues.

```
509  pico
867  pico
620  bagels
981  pico
794  bagels
???
```

2. Solve.

What is the probability of rolling a 6 on one or the other of a pair of 6-sided dice or a sum of 6 on the two dice?

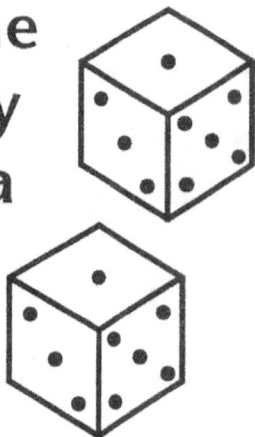

3. Sketch.

If it takes Igor 6 minutes to saw a log into 3 pieces. How long would it take him to saw an equivalent log into 4 pieces?

4. Review.

MATHERCISE E27

1. Reason.

Each of the three letters in the sum on the right represents a different digit. What is the value of B?

```
    A  O  5
    8  C  A
    B  A  7
 +  7  8  8
 ───────────
 C  3  8  6
```

2. Solve.

A wire of uniform thickness and density weighs 48 kilograms. It is cut into 2 pieces. One piece is 45 meters long and weighs 36 kilograms. What is the length in meters of the other piece?

3. Sketch.

Sketch the graph of $y = x^2 - 2$.

4. Review.

MATHERCISE E28

1. Reason.

From the given set of clues, find a 3-digit number that satisfies all the clues.

```
567  pico
867  pico  pico
620  pico  pico
086  fermi  fermi
???
```

2. Solve.

$$x = 12y$$
$$8y = 6w$$
$$4x = 18z$$

Find z in terms of w.

3. Sketch.

Sketch the graph of $y = (x + 5)^2$.

4. Review.

MATHERCISE E29

1. Reason.

Each of the three letters in the difference on the right represents a different digit. What is the value of C?

$$
\begin{array}{r}
A\ 5\ B \\
-\ 1\ A\ 5 \\
\hline
C\ 8\ 8
\end{array}
$$

2. Solve.

$x = 3y$

$y = 10w$

$z = 3w$

Find z in terms of x.

3. Sketch.

If the graph is $y = f(x)$, sketch the graph of $y = f(x) - 3$.

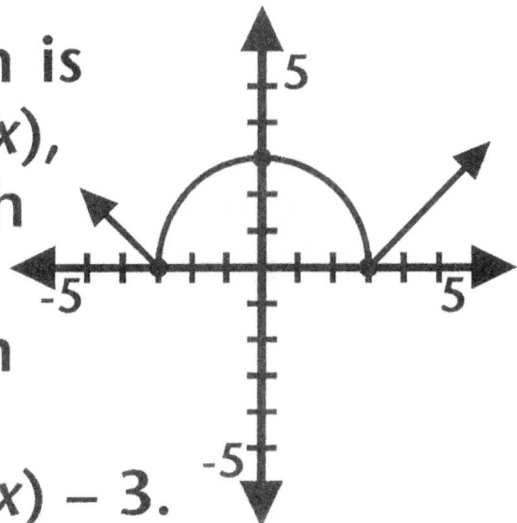

4. Review.

MATHERCISE E30

1. Reason.

From the given set of clues, find a 3-digit number that satisfies all the clues.

```
509  pico
867  pico
620  pico
281  pico  fermi
437  bagels
258  pico  pico
???
```

2. Solve.

$$f(x) = x^2 + 3x + 2$$

Find $f(x + h)$.

3. Sketch.

Sketch one period of the graph of $y = \sin x + 2$.

4. Review.

MATHERCISE E31

1. Reason.

What comes next in the pattern?

0, 1, 6, 31, 156, 781, __?__ ,

2. Solve.

$f(x) = x^2 + 3x + 2$

Find $f(x + h) - f(x)$.

3. Sketch.

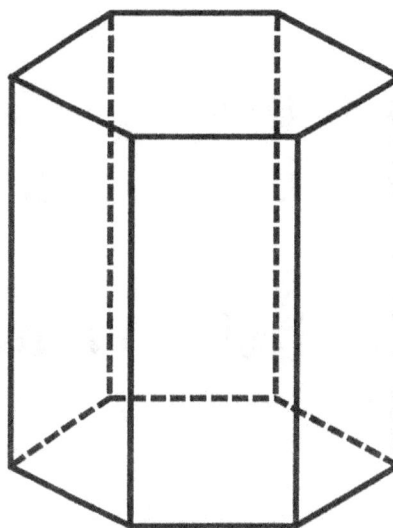

4. Review.

MATHERCISE E32

1. Reason.

If the pattern continues to the right, draw how the 17th box would look.

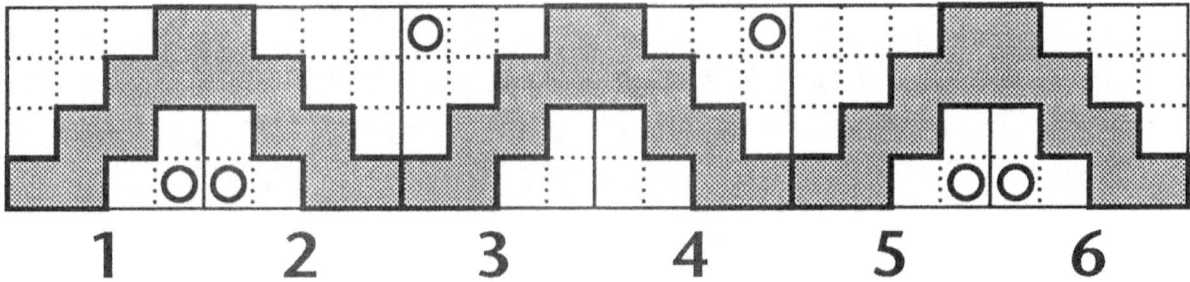

1   2   3   4   5   6

2. Solve.

$a \blacklozenge b = 3a + b$
$a * b = a - 3b$

Find the value of $6 \blacklozenge (2 * 4)$.

3. Sketch.

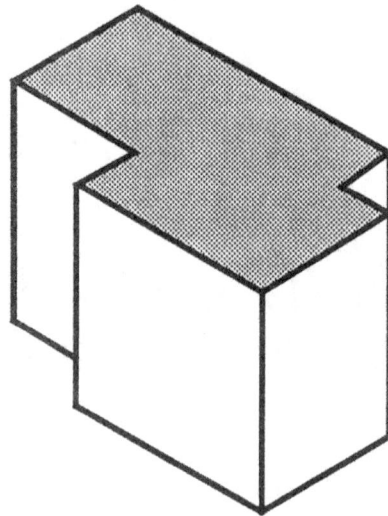

4. Review.

MATHERCISE E33

1. Reason.

What comes next in the pattern?

7, 8, 5, 10, 30, 31, 28, 33, 99, 100, 97, 102, 306, ___?___ ,

2. Solve.

$a * b = a^2 - ab + b^2$

$(k * 3) = 19$

Find k.

3. Sketch.

Sketch this figure resting on its shaded faces.

4. Review.

MATHERCISE E34

1. Reason.

If the pattern continues to the right, draw how the 19th box would look.

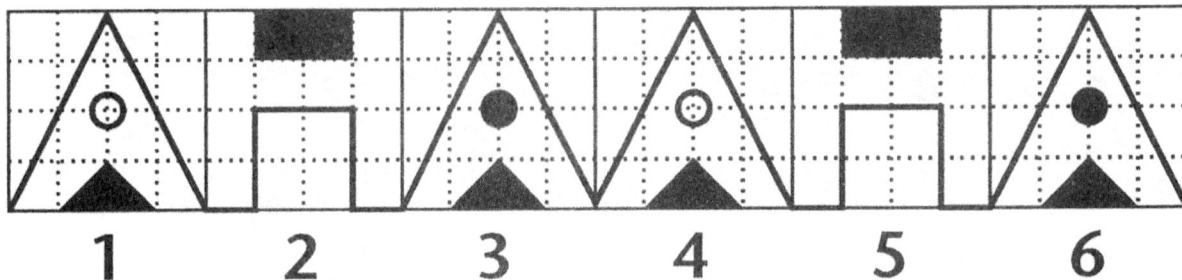

2. Solve.

What is the length of the median *AM* if *M* is the midpoint of side *BC* in triangle *ABC* with vertices *A* at (0, 0), *B* at (12, 0), and *C* at (8, 6)?

3. Sketch.

Sketch what this shape would look like if it were folded along the dotted lines into a solid figure.

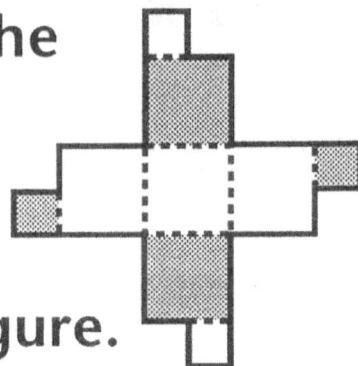

4. Review.

MATHERCISE E35

1. Reason.

I am facing the front of my house having an early breakfast. While looking out my front window I see a woman walking by the front of the house, her shadow directly in front of her, and with my house at her right. Which direction am I facing?

2. Solve.

What is the equation of the line through the median AM if M is the midpoint of side BC in triangle ABC with vertices A at $(0, 0)$, B at $(12,0)$, and C at $(8, 6)$?

3. Sketch.

Are the two shapes that make up the rectangle identical? (Ignoring the shading!)

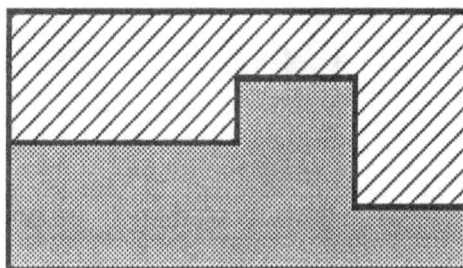

4. Review.

MATHERCISE E36

1. Reason.

If deleting all occurances of the letter "*t*" from the word *battle* leaves a meaningful four letter word, print that word. Otherwise, print the word *war* in a vertical column with the last letter on top and the first letter at the bottom and circle the middle letter.

2. Solve.

What is the vertex of the parabola $y = x^2 + 3$?

3. Sketch.

The shape shown is half of a 6x6 square. Sketch a 6x6 square showing the two halves identical to the one shown.

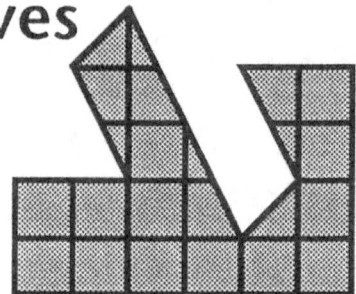

4. Review.

MATHERCISE E37

1. Reason.

Breakfast cereal Whako Wheaties has 5 more milligrams of salt than Koko Crispies and 20 more milligrams of salt than Special Flakes. Fruity Flakes has 15 milligrams less salt than Special Flakes and 25 milligrams less salt than Corn Toasties. If Corn Toasties has 240 milligrams of salt per serving, then how many milligrams of salt does Whako Wheaties have per serving?

2. Solve.

What is the probability of rolling a 2 on one or the other of a pair of 6-sided dice or a sum of 2 on the two dice?

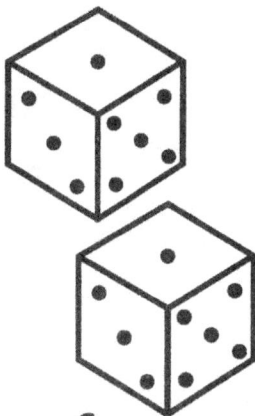

3. Sketch.

Sketch the graph of $y = (x + 2)^2 - 4$.

4. Review.

MATHERCISE E38

1. Reason.

I am facing the front of my house having an early breakfast. While looking out my front window I see a woman walking by the front of the house, her shadow directly behind her, and with my house at her left. Which direction am I facing?

2. Solve.

A bag contains 4 red, 3 blue, and 2 green marbles. Two marbles are taken from the bag. What is the probability that the 2 marbles are red?

3. Sketch.

Sketch the graph of $y = x^2 - 4x + 7$.

4. Review.

MATHERCISE E39

1. Reason.

Steve, Mark, and Robert are married to Susan, Martha, and Ruby, but not necessarily in that order. Mark's children have never played with Ruby's daughter. Ruby's daughter is in the same class with one of Susan's children. Robert has three children, all boys. Martha, who is Robert's sister, has no children. Match up the couples.

2. Solve.

The head waiter at *Chez Pansy* Restaurant recorded that for every 21 customers that asked for the smoking section, 84 requested non-smoking. If the seating area is about 500 square feet, about how many square feet will be for non-smokers?

3. Sketch.

If the graph is $y = f(x)$, sketch the graph of $y = f(x) + 4.$

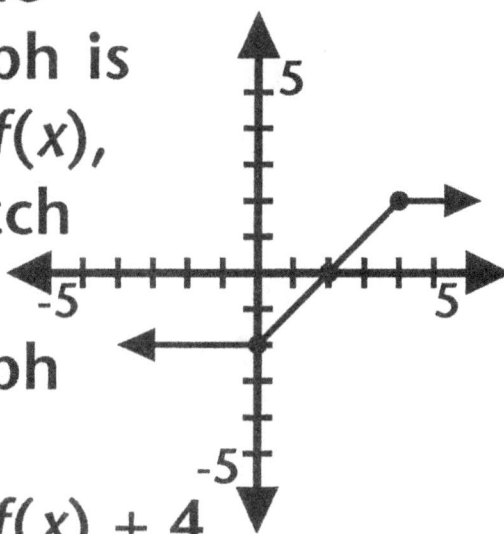

4. Review.

MATHERCISE E40

1. Reason.

Miguel and Rosita are the same height. Miguel is heavier than Donna. Donna is shorter than Nathan. Rosita is taller than Donna but weighs the same as Nathan. Nathan is lighter than Miguel but heavier than Donna and taller than Rosita. Is it possible to rank Miguel and Rosita by weight? If yes, who is the lightest of the two?

2. Solve.

$B = \frac{1}{2} h(a + b)$

$V = \frac{1}{3} BH$

$h = 2x$

$a + b = 3x$

$Hx = 12$

Find V in terms of x.

3. Sketch.

Sketch one period of the graph of $y = 2 \cos x - 3$.

4. Review.

MATHERCISE E41

1. Reason.

What comes next in the pattern?

-6, 4, 20, 42, 70, 104, 144, 190, 242, ___?___ ,

2. Solve.

$A = \frac{1}{2} bh$

$V = \frac{1}{3} pA$

$b = 2y$

$h = \frac{1}{2} b$

$p = 30$

Find V in terms of y.

3. Sketch.

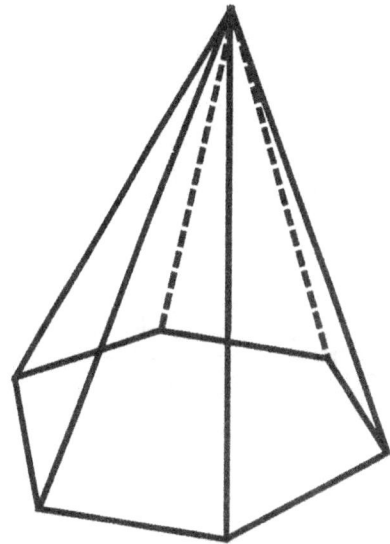

4. Review.

MATHERCISE E42

1. Reason.

If the pattern continues to the right, draw how the 23rd box would look.

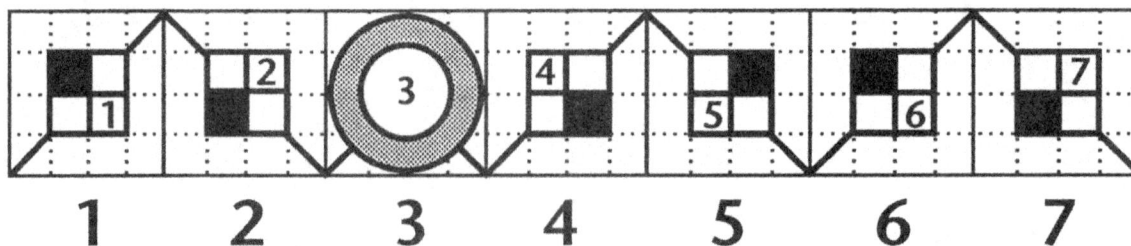

2. Solve.

$$f(x) = x^2 + 3x + 2$$

Find $\dfrac{f(x + h) - f(x)}{h}$.

3. Sketch.

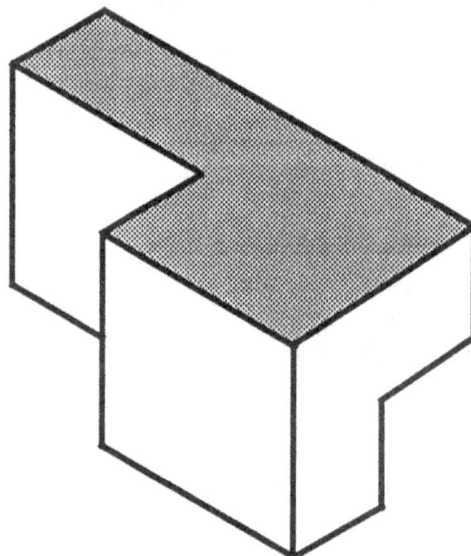

4. Review.

MATHERCISE E43

1. Reason.

What comes next in the pattern?

14, 45, 88, 143, 210, 289, 380, 483, ___?___ ,

2. Solve.

$f(x) = x^3 + 2$

Find $\dfrac{f(x + h) - f(x)}{h}$.

3. Sketch.

A rectangular sheet of paper is folded in half vertically, then the two halves folded in half horizontally so that the 4 corners come together. Snip off a right isosceles triangle from that corner. Sketch what the paper will look like when unfolded.

4. Review.

MATHERCISE E44

1. Reason.

If the pattern continues to the right, draw how the 99th box would look.

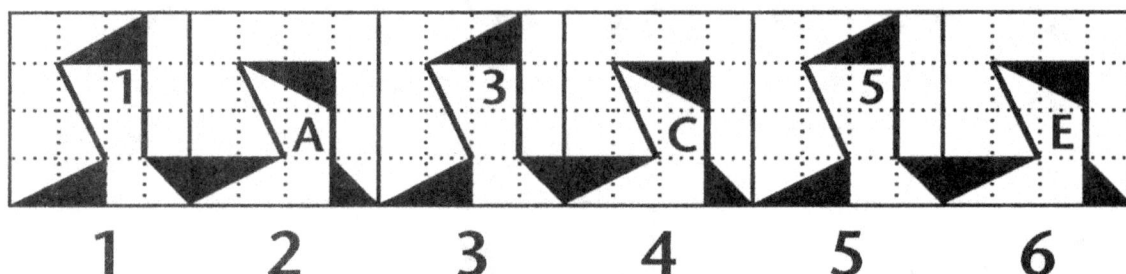

2. Solve.

Let

$a \blacklozenge b = ab + a + b$.

For what value of x is the statement $x \blacklozenge y = x$ always true?

3. Sketch.

The shape shown is half of a 6x6 square. Sketch a 6x6 square showing the two halves identical to the one here.

4. Review.

MATHERCISE E45

1. Reason.

Race car drivers Tom, Dick, and Harriet took first, second, and third but not necessarily in that order in the Indiana Jones Memorial Sports Car Race. The top three finishers drove a Porsche, Alfa Romeo, and Corvette but not necessarily in that order. Dick placed higher than the Alfa but lower than Harriet. The Corvette took first. Match each driver with their car and final position in the race.

2. Solve.

$\Delta x = \frac{x}{2}$ if x is even.

$\Delta x = 2x$ if x is odd.

What is $\Delta(4Q + 1)$ when Q is a positive integer?

3. Sketch.

Are the two shapes that make up the rectangle identical? (Ignoring the shading!)

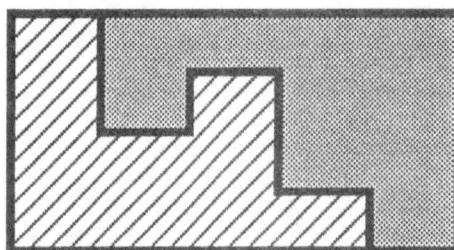

4. Review.

Mathercise Book E

MATHERCISE E46

1. Reason.

Each of the 3 letters in the sum on the right represents a different digit. What is the value of $A + B + C$?

$$\begin{array}{r} A \ B \ A \\ + \ C \ B \ C \\ \hline 4 \ 4 \ 4 \end{array}$$

2. Solve.

What is the equation of the axis of symmetry of the parabola $y = (x + 1)^2 - 3$?

3. Sketch.

A rectangular sheet of paper is folded in half vertically, then the two halves folded in half horizontally so that the 4 corners come together. Snip off a right isosceles triangle from all four corners. Sketch what the paper will look like when unfolded.

4. Review.

MATHERCISE E47

1. Reason.

From the given set of clues, find a 3-digit number that satisfies all the clues.

```
094  pico
132  pico
379  pico
716  pico  fermi
582  bagels
701  pico  pico
???
```

2. Solve.

What is the equation of the parabola that has its only y-intercept at (0, 8) and a vertex of (2, 4)?

3. Sketch.

If the graph is $y = f(x)$, sketch the graph of $y = f(x + 3) + 2$.

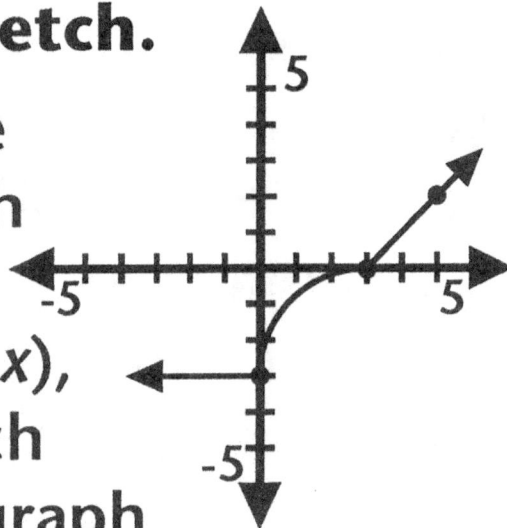

4. Review.

MATHERCISE E48

1. Reason.

Jockies Toni, Doreen, and Natalie took first, second, and third but not necessarily in that order in the Indiana Jane Memorial Horse Race. The top three finishers rode a black stallion, a brown mare, and spotted gelding but not necessarily in that order. Doreen placed higher than the brown mare but lower than Natalie. The spotted gelding took first. Match each rider with her horse and final position in the race.

2. Solve.

What is the equation of the parabola that has one of its two x-intercepts at (-5, 0), and a vertex of (-1, 2)?

3. Sketch.

Sketch one period of the graph of $y = \sin\left(x - \frac{\pi}{2}\right)$.

4. Review.

MATHERCISE E49

1. Reason.

Each of the 5 letters in the product on the right represents a different digit. What is the value of *A*?

```
        A  B  5  C
   X          5  D
   ─────────────────
        A  B  5  C
   D  7  2  8  E
   ─────────────────
   D  7  C  2  5  C
```

2. Solve.

A bag contains 4 red, 3 blue, and 2 green marbles. Two marbles are taken from the bag. What is the probability that the 2 marbles are blue?

3. Sketch.

If the graph below is the graph of $y = f(x)$, sketch the graph of $y = f(x + 3)$.

4. Review.

MATHERCISE 50

1. Reason.

From the given set of clues, find a 3-digit number that satisfies all the clues.

```
674  pico
032  pico
519  fermi
349  bagels
560  pico
???
```

2. Solve.

A bag contains 4 red, 3 blue, and 2 green marbles. Two marbles are taken from the bag. What is the probability that the 2 marbles are green?

3. Sketch.

Sketch one period of the graph of
$y = \cos\left(x - \frac{\pi}{4}\right) + 1.$

4. Review.